NATURE'S MYSTERIES

ECLIPSES

CORONA BREZINA

Britannica Educational Publishing
IN ASSOCIATION WITH
ROSEN
EDUCATIONAL SERVICES

Published in 2019 by Britannica Educational Publishing (a trademark of Encyclopædia Britannica, Inc.) in association with
The Rosen Publishing Group, Inc.
29 East 21st Street, New York, NY 10010

Distributed exclusively by Rosen Publishing.
To see additional Britannica Educational Publishing titles, go to rosenpublishing.com.

First Edition

Britannica Educational Publishing
J.E. Luebering: Executive Director, Core Editorial
Mary Rose McCudden: Editor, Britannica Student Encyclopedia

Rosen Publishing
Kathy Kuhtz Campbell: Senior Editor
Michael Moy: Series Designer
Tahara Anderson: Book Layout
Cindy Reiman: Photography Manager
Nicole DiMella: Photo Researcher

Library of Congress Cataloging-in-Publication Data

Names: Brezina, Corona, author.
Title: Eclipses / Corona Brezina.
Description: First edition. | New York: Britannica Educational Publishing, in Association with Rosen Educational Services, 2019. | Series: Nature's mysteries | Includes bibliographical references and index. | Audience: Grades 1–5.
Identifiers: LCCN 2018007401| ISBN 9781508106616 (library bound) | ISBN 9781508106494 (pbk.) | ISBN 9781508106555 (6 pack)
Subjects: LCSH: Eclipses—Juvenile literature. | Solar eclipses—Juvenile literature. | Lunar eclipses—Juvenile literature.
Classification: LCC QB175 .B74 2019 | DDC 523.9/9—dc23
LC record available at https://lccn.loc.gov/2018007401

Manufactured in the United States of America

Photo credits: Cover, p. 1 and interior pages background muratart/Shutterstock.com; pp. 4, 11, 14, 22, 23, 27 NASA; pp. 5, 17 © Encyclopædia Britannica, Inc.; p. 6 Classic Image/Alamy Stock Photo; p. 7 Print Collector/Hulton Archive/Getty Images; p. 8 Kyodo News/Getty Images; p. 9 NASA/Hinode/XRT; p. 10 NASA/Aubrey Gemignani; pp. 12, 25 The Washington Post/Getty Images; p. 13 GeneBlevins/AFP/Getty Images; p. 15 Mandel Ngan/AFP/Getty Images; p. 16 ChameleonsEye/Shutterstock.com; p. 18 Allen J. Schaben/Los Angeles Times/Getty Images; p. 19 Xinhua News Agency/Getty Images; p. 20 Serge Koutchmy, IAP (CNRS), SOHO (ESA&NASA); p. 21 Roger Viollet/Getty Images; p. 24 ClassicStock/Alamy Stock Photo; p. 26 Frank Zullo/Science Source; p. 28 Chris Butler/Science Source; p. 29 NASA/Ames/JPL-Caltech.

CONTENTS

WHAT ARE ECLIPSES?

An eclipse happens when one object in space blocks another from view. Eclipses happen because planets, moons, and other **celestial bodies** constantly move through space. As Earth moves around the sun in a path called an orbit, the moon makes its own orbit around Earth. At certain points in each orbit, the sun, moon, and Earth line up in a row.

This artwork shows how the sun, moon, and Earth align to create an eclipse of the sun as seen on Earth.

During a solar eclipse the moon comes between Earth and the sun for a short time. The moon's shadow falls on Earth's surface. The inner part

VOCABULARY

Celestial bodies are objects located in the sky, such as the moon, other planets, asteroids, and stars.

of the shadow is called the umbra. People watching from within the umbra experience a total solar eclipse. In a total solar eclipse, the moon appears as almost exactly the right size to completely cover the sun when observed from Earth. The outer part of the shadow is the penumbra. People viewing from within the penumbra see a partial eclipse.

Another type of eclipse happens when Earth comes between the moon and the sun. That is called a lunar eclipse.

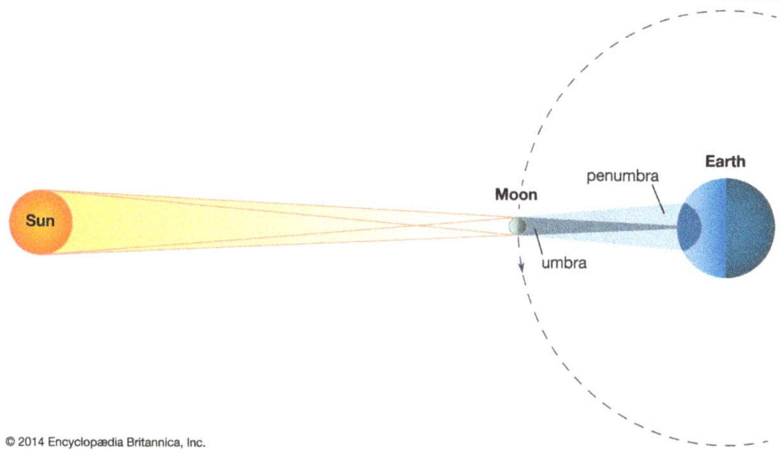

In the darkly shaded region (umbra) a solar eclipse is total; in the lightly shaded region (penumbra) the eclipse is partial.

ANCIENT ECLIPSES

Some ancient peoples regarded total solar eclipses as omens—signs that a disaster was coming. They were terrified when the sky went dark in the middle of the day for no reason they could understand.

The first known record of an eclipse dates back to March 3, 1223 BCE. The eclipse was seen in the city of Ugarit in what is now Syria. Most ancient records come from Greece, China, and

To ancient peoples, eclipses were sometimes seen as signs that the gods were angry.

THINK ABOUT IT

Modern astronomers can accurately calculate the dates of past eclipses. Many ancient records do not include exact dates of eclipses. Why might a historian be excited to find a mention of an eclipse in a historic document?

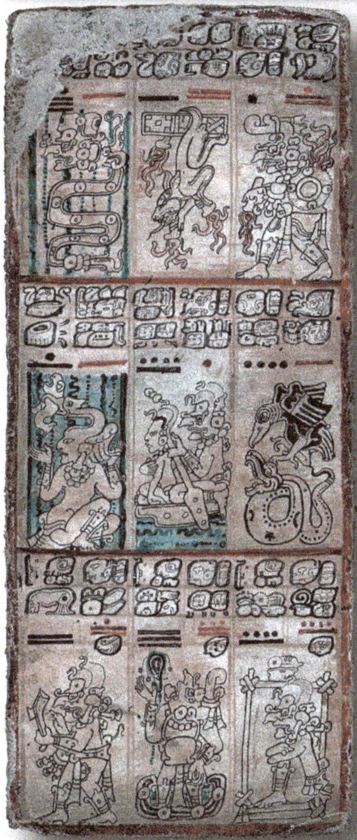

Babylonia (a region in what is now Iraq). Astronomers in Baghdad and Cairo carefully recorded many eclipses between about 800 and 1000 CE. In about 1450, European astronomers started making fairly accurate measurements of the time of day or night when eclipses occurred.

Eclipses are mentioned in ancient works of history, literature, and science. Today, these records are useful to both astronomers and historians.

This page is from a Mayan text written about eight hundred years ago. It includes tables of solar and lunar eclipses.

SOLAR ECLIPSES

There are different kinds of solar eclipses. When the moon covers only part of the sun, a partial solar eclipse occurs. A total solar eclipse takes place when the moon passes across the sun's disk so that the sun appears completely covered by the moon. All viewers can see during a total eclipse is a ring of light called a corona. The corona is part of the sun's atmosphere. The darkness of a total eclipse lasts for only a few minutes.

Sometimes, the moon is too far from Earth to appear big

Photographs of a solar eclipse capture the moon moving across the face of the sun.

COMPARE AND CONTRAST

What is the difference between an annular solar eclipse and a partial solar eclipse?

A satellite image of an annular eclipse shows the ring of fire created by the moon partially covering the sun.

enough to totally cover the sun. If the moon passes directly across the sun's disk at this time, the moon does not fully cover it. The rim of the sun is still visible around the edge of the moon. This is called an annular eclipse.

OBSERVATIONS OF A TOTAL SOLAR ECLIPSE

A total solar eclipse begins when the disk of the moon appears to touch the disk of the sun. This is called "first contact." The moon gradually moves to cover more of the face of the sun. The sun shrinks to a smaller and smaller crescent. The light starts to dim.

A few seconds before totality—the point

Baily's beads (*lower left*) can be viewed just before totality during a solar eclipse.

THINK ABOUT IT

The temperature changes by about 10 degrees Fahrenheit (6 degrees Celsius) during a total solar eclipse. Do you think it gets hotter or colder?

at which the moon covers the sun completely—viewers can see "Baily's beads." These are bright specks around the edge of the moon. They are created by sunlight shining through valleys at the moon's edge.

Totality begins when the moon completely blocks the sun. This is called "second contact." Observers can see the sun's corona. The light from the corona is much fainter than sunlight. Pearly white coronal streamers can be seen far beyond the moon's dark disk.

The sun's corona, its thin outer atmosphere, shines brightly around the moon during a total solar eclipse.

Viewers can observe many strange changes in the world around them throughout a total solar eclipse. During the partial stages, sunlight shining through gaps in tree leaves and other small openings forms little crescents of light on the ground. Animals are sometimes scared, or they might get ready to go to sleep as if it is dusk. In the minutes before and after totality, "shadow bands" of rippling light can be seen on light-colored surfaces. They are caused by **refraction** of sunlight by disturbances in

During the partial stages of a solar eclipse, light from the sun forms crescent-shaped shadows on the ground and other surfaces.

VOCABULARY

Refraction happens when light changes direction, or bends, as it moves from one material to another.

Earth's atmosphere. At totality, planets and stars are visible in the dark sky.

Totality ends with "third contact" as the moon begins to move beyond the sun. Baily's beads appear once again. The crescent of the sun starts to grow bigger. The eclipse ends with "fourth contact" when the moon no longer covers any part of the sun.

As the sun emerges from totality, rays shine like a jewel on one edge, creating a diamond ring effect.

VIEWING A TOTAL SOLAR ECLIPSE

Total solar eclipses are rare and spectacular events. They are only visible across a thin path on Earth's surface. People sometimes travel long distances to witness a total eclipse. Partial and annular eclipses also attract stargazers.

Astronomers have created maps and timetables that can be help-ful for observers. An eclipse occurs over the period of a couple hours. Total-ity lasts for a couple

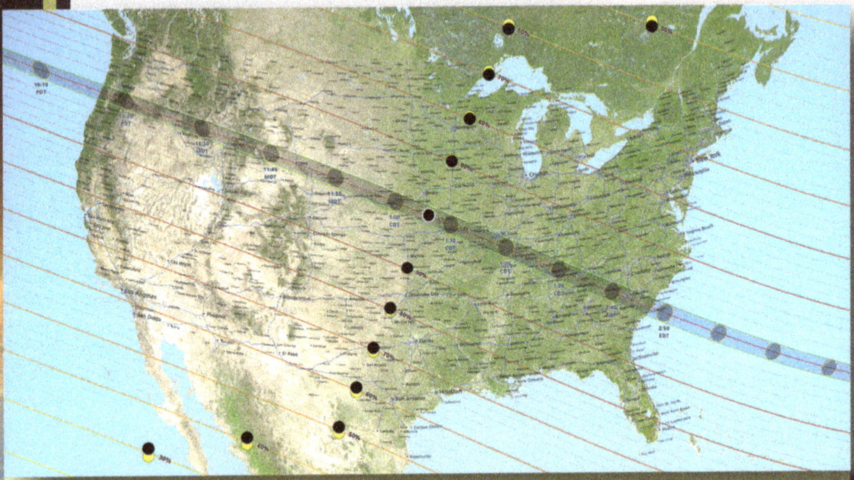

A solar eclipse map shows the path of the moon's shadow and the times viewers will experience the total eclipse.

THINK ABOUT IT

Places such as observatories and science museums may hold eclipse events. People can learn about eclipses and perhaps view the eclipse through a telescope with a solar filter. Where could you go to view an eclipse or other astronomical event near you?

of minutes. The times for the points of contact vary depending on an observer's specific location.

Viewers safely watch a total solar eclipse through eclipse glasses.

During an eclipse, an observer should never look directly at the sun, even if it is partially covered by the moon. Filtering devices such as eclipse glasses can allow safe viewing of a solar eclipse. Totality is the only period when an observer can look at an eclipse directly.

LUNAR ECLIPSES

The moon does not make any light itself. Rather, the sun shines on the moon, and the reflected sunlight makes the moon visible from Earth.

During a lunar eclipse, Earth blocks sunlight from reaching the moon. The sun, Earth, and the moon line up, with Earth in the middle. The moon is then in Earth's shadow. But the moon does not appear completely dark. Instead, it glows a dim orange or red color. This is because some light reaches the moon indirectly.

The moon slowly dims during a lunar eclipse until the entire face is covered by Earth's shadow.

THINK ABOUT IT

Why do you think a totally eclipsed moon is sometimes called a blood moon?

This light reaches the moon after being refracted by Earth's atmosphere. The temperature of the moon's surface drops during a lunar eclipse.

A total lunar eclipse occurs when the whole moon passes through the center of Earth's umbra. A partial lunar eclipse is observable when only a part of the moon moves through the umbra. A penumbral eclipse occurs when the moon passes only through Earth's penumbra.

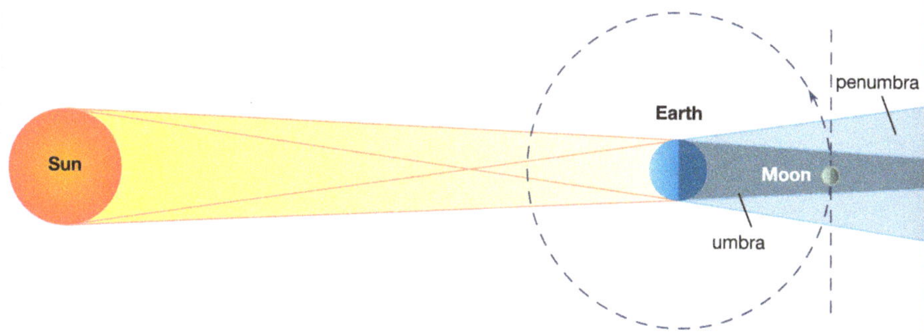

This diagram shows a lunar eclipse. The umbra is the total shadow. The penumbra is the partial shadow.

VIEWING A LUNAR ECLIPSE

Lunar eclipses generally occur twice every year. They always happen during a full moon. Unlike solar eclipses, lunar eclipses can be seen anywhere the moon is visible.

A total eclipse of a supermoon can be quite spectacular to see. A supermoon is a full moon that occurs when the moon is at the closest point to Earth in its orbit. An eclipse of this type occurred on January 31, 2018.

A lunar eclipse occurs over a longer period of time than a solar eclipse does. A total lunar eclipse begins when the Earth's umbra falls on the edge of the moon. The moon's surface gradually becomes darkened. After

COMPARE AND CONTRAST

How are lunar eclipses similar to solar eclipses? How are they different?

about an hour, when the whole disk of the moon is within the umbra, the eclipse becomes total. If the moon's path leads through the center of the umbra, totality can last longer than an hour and a half.

A series of photographs shows Earth's shadow passing over the moon during a lunar eclipse, ending in a totality.

Eye protection is not necessary during a lunar eclipse. Viewing the eclipse through binoculars or a telescope can reveal details of the moon's surface.

ECLIPSE RESEARCH ACTIVITIES

Eclipses give scientists a chance to learn more about the sun. Solar eclipses have provided astronomers with a chance to observe the sun's faint outer atmosphere. In the 1930s, for instance, German astronomer Walter Grotrian concluded that the sun's corona was unbelievably hot based on information he gathered from a total eclipse.

Another notable discovery during a solar eclipse involved helium. Helium is a gas that has no color, odor, or taste. In 1868, French astronomer Pierre Janssen predicted the existence of the gas based on

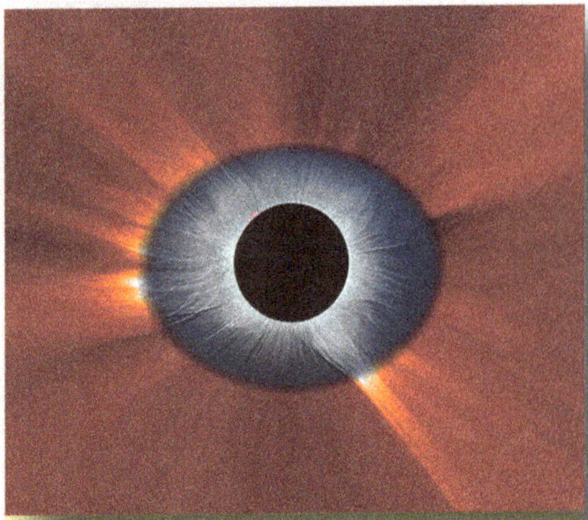

The sun's corona is seen in an eclipse image that was combined from ground and satellite views.

VOCABULARY

A theory is an idea or set of ideas that is intended to explain facts or events.

observations of the corona. The word "helium" comes from the Greek word for sun. Helium was not found on Earth until 1895.

A measurement taken during a total eclipse of the sun in 1919 even supported Albert Einstein's general **theory** of relativity. According to the theory, light from other stars would behave in a certain way near the sun during an eclipse. Experiments proved that Einstein was correct.

Scientists also perform research during lunar eclipses. These eclipses are the only times that they can observe the effects of the temperature dropping quickly on the moon. Scientists can learn more about the properties of the surface of the moon.

Albert Einstein's theories changed scientific understanding of the nature of gravity, time, and space.

PREDICTING ECLIPSES

Eclipses are rare occurrences because the moon travels around Earth in a slightly tilted orbit. Scientists use mathematics and the laws of physics to predict the motion of the sun, moon, and Earth. Eclipses occur in regular cycles. If a solar eclipse occurs, a similar eclipse will happen about eighteen years, eleven days, and eight hours later. This is because the sun, moon, and Earth return to almost the same position. The cycle may continue for centuries.

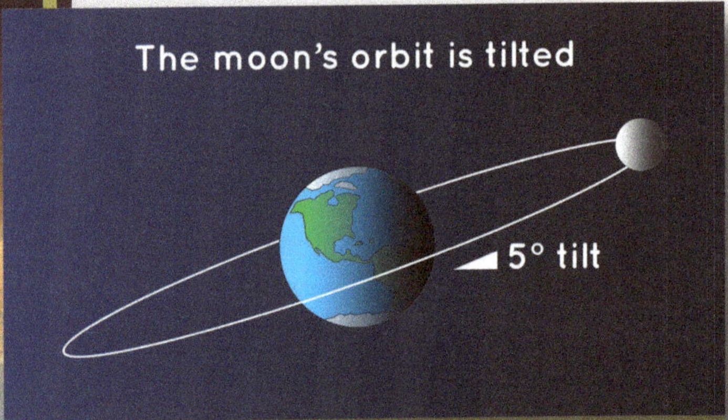

Lunar eclipses occur infrequently because the moon travels around Earth in a path that is slightly tilted.

There are dozens of these cycles in progress. As a result, most

THINK ABOUT IT

Earth's surface features such as mountains can affect whether a location will fall under the moon's shadow during a solar eclipse. Why might that be?

years have two to five solar eclipses. Similar calculations are used to predict lunar eclipse cycles.

Scientists use highly detailed maps of both Earth and the moon to determine where solar eclipses will be visible. The eclipses can be calculated decades in advance.

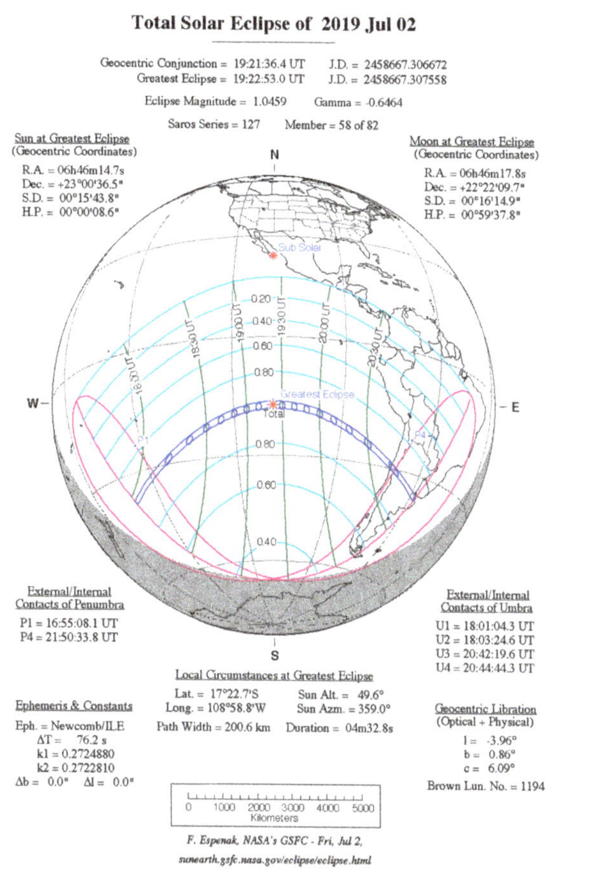

Scientists can calculate an eclipse's path, timing, and type (total, annular, or partial) many years in advance.

ECLIPSES IN THE UNITED STATES

Americans have long been dazzled by total solar eclipses. Thomas Jefferson was interested in eclipses and tried to view several. On one occasion, he hoped to observe an eclipse on June 24, 1778, from his home in Virginia. He was disappointed when the day was cloudy. On June 16, 1806, John Quincy Adams viewed a total solar eclipse from Boston. He observed several other eclipses and later described one in a poem.

In 1878, tourists flocked to the center of the country for an

HARPER'S WEEKLY
A JOURNAL OF CIVILIZATION

A magazine cover shows tourists viewing the "great" solar eclipse of 1878 in Colorado, in the Rocky Mountains.

THINK ABOUT IT

Explorer Christopher Columbus scared native people of what is now Jamaica by telling them that the full moon would turn an angry red on February 29, 1504. What astronomical event did he predict?

eclipse on July 29. Thomas Edison watched it from Wyoming. A century later, an eclipse on February 26, 1979, was visible in several northwestern states.

An eclipse on August 21, 2017, was visible across the center of the United States. The path of totality passed over twelve different states. The next total eclipse visible in the United States will occur on April 8, 2024.

Millions of Americans, including these students in Washington, DC, viewed the "Great American Eclipse" of August 21, 2017.

OTHER TYPES OF ECLIPSES

The moon (*top*) passed in front of Jupiter (*top, right*) in an occultation and briefly blocked the planet from view. Venus can be seen at the lower left.

Eclipses can involve celestial bodies besides Earth, the sun, and the moon. There are specialized terms for these events. An occultation occurs when one celestial body passes in front of what looks like a much smaller one. An example is a planet passing in front of a star, hiding the star from a viewer on Earth. Observing occultations can help astronomers learn about the sizes and shapes of planets and faraway stars.

COMPARE AND CONTRAST

How is an occultation similar to a solar eclipse? How is it different?

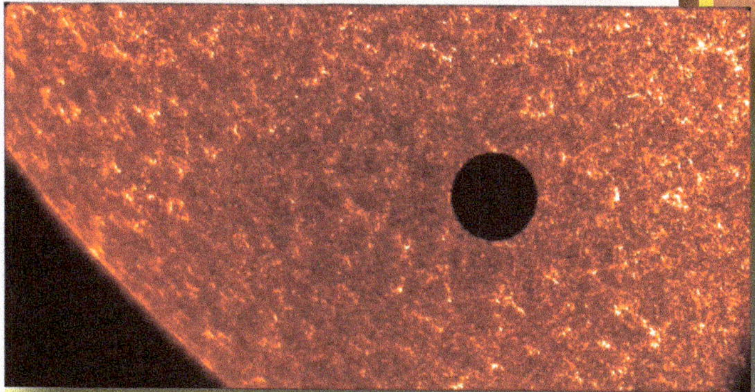

A satellite orbiting Earth captured this image of the planet Venus transiting the sun.

A transit occurs when a small celestial body moves across the face of a much larger one. The planets Mercury and Venus sometimes transit the sun as viewed from Earth. Mercury transits the sun about thirteen times per century. Transits of Venus are much rarer. Observations of the transit of Venus provided valuable scientific information in the eighteenth and nineteenth centuries. Astronomers made measurements that allowed them to calculate the distance between Venus and Earth. This distance allowed them to figure out the distance between the sun and Earth.

BEYOND THE SOLAR SYSTEM

Eclipses can occur beyond the solar system. Some stars belong to systems of multiple stars. A kind of double star called a binary star is made up of two stars that orbit each other. In eclipsing binary stars, two eclipses take place during each orbital period, or revolution. The amount of light given off by the pair of stars varies as one star passes in front

The light shining from an eclipsing binary star dims when one star passes in front of the other.

VOCABULARY

Exoplanets are planets that orbit stars that are beyond the solar system. They are also called extrasolar planets.

of or behind the other. Using information gathered during these eclipses, astronomers can calculate the stars' properties such as sizes, shapes, and temperatures.

In 2010, two exoplanets were observed transiting the same star. It was the first discovery of a system with multiple planets.

Astronomers discovered many **exoplanets** after they observed them transiting their stars. The brightness of a star drops as the planet moves across the star's face. If a star dims and brightens in a regular pattern, the effect is likely caused by a planet. Sometimes, multiple planets have been discovered orbiting the same star.

GLOSSARY

ATMOSPHERE The whole mass of air surrounding Earth; also, a mass of gases surrounding a heavenly body.

CALCULATE To find by performing mathematical operations (such as addition, subtraction, multiplication, and division).

CORONA The outermost part of the atmosphere of a star (such as the sun).

CRESCENT The curved shape of the moon when it is less than half full.

CYCLE A period of time taken up by a series of events or actions that repeat themselves regularly and in the same order.

FILTER A transparent material (such as colored glass) that absorbs light of some wavelengths and is used to change light.

GRAVITY A pulling force between any two objects that works across space. The greater the mass of an object, the greater is its force of gravity.

HELIUM A chemical element that is a colorless gas that is lighter than air.

LUNAR Of, relating to, or resembling the moon.

OBSERVE To gather information (such as for scientific studies) by noting facts or occurrences; to watch.

OMEN A happening believed to be a sign or warning of some future event.

ORBIT The path taken by one body circling around another body, for example, the orbit of Earth around the sun; also, one complete circle that makes up such a path.

RECORD Something that recalls or tells about past events.

REVOLUTION The time taken for a heavenly body to complete one orbit.

SATELLITE A man-made object or vehicle orbiting the earth, the moon, or another heavenly body.

SOLAR Of, derived from, relating to, or caused by the sun.

VISIBLE Capable of being seen.

FOR MORE INFORMATION

Books

Bingham, Caroline, and Ishani Nandi. *First Space Encyclopedia*. Rev. ed. New York, NY: DK Publishing, 2016.

Fraknoi, Andrew, and Dennis Schatz. *When the Sun Goes Dark*. Arlington, VA: National Science Teachers Association, 2017.

Owen, Ruth. *Solar and Lunar Eclipses* (Explore Outer Space). New York, NY: PowerKids Press, 2013.

Rustad, Martha E. H. *Eclipses*. North Mankato, MN: Capstone Press, 2018.

Sherman, Jill. *Exploring Eclipses* (Fact Finders: Discover the Night Sky). Mankato, MN: Capstone Press, 2018.

Websites

NASA Space Place

https://spaceplace.nasa.gov/eclipse-snap/en/

National Air and Space Museum

https://airandspace.si.edu/

Facebook, Twitter: @airandspace

Instagram: @airandspacemuseum

NeoK12

https://www.neok12.com/Eclipse.htm

PBS LearningMedia

https://ny.pbslearningmedia.org/resource/ess05.sci.ess.eiu.eclipse/solar-eclipses/#.WoMSKmaZO-o

Rader's Cosmos4Kids

http://www.cosmos4kids.com/files/systems_eclipses.html

INDEX